Immunotherapy 2018

Topical Conference at the 2018 AIChE Annual Meeting

Pittsburgh, Pennsylvania, USA
28 October - 2 November 2018

ISBN: 978-1-5108-7645-3

Printed from e-media with permission by:

Curran Associates, Inc.
57 Morehouse Lane
Red Hook, NY 12571

Some format issues inherent in the e-media version may also appear in this print version.

Printed by Curran Associates, Inc. (2019)

For permission requests, please contact AIChE
at the address below.

AIChE
120 Wall Street, FL 23
New York, NY 10005-4020

Phone: (800) 242-4363
Fax: (203) 775-5177

www.aiche.org

Additional copies of this publication are available from:

Curran Associates, Inc.
57 Morehouse Lane
Red Hook, NY 12571 USA
Phone: 845-758-0400
Fax: 845-758-2633
Email: curran@proceedings.com
Web: www.proceedings.com

TABLE OF CONTENTS

Some abstracts in this book refer to images that were not made available by the publisher

539452 Eradication of Primary and Cancer Stem Cells By Chemically Self-Assembled Nanoring Targeted T-Cells

Abstract:

Over the past decade, it has become increasingly clear that harnessing a cancer patientâ€™s T-cells to destroy tumor cells will be an important weapon in an oncologistâ€™s therapeutic arsenal. Nevertheless, although selective tumor cell killing and significant clinical success against B-cell malignancies has been demonstrated, the same level of success against solid tumors has remained elusive.

To address this unmet need, our group has envisioned the potential of chemical biology and protein engineering to non-genetically modify T-cell membranes. We have shown that two dihydrofolate reductase molecules ($DHFR^2$) fused to a single chain antibody (scFv) or peptide can spontaneously self-assemble upon the addition of the chemical dimerizer, bis-methotrexate (BisMTX), into highly stable multivalent targeted **chemically self-assembled nanorings (CSANs)**. Recently, we have prepared anti-EpCAM/anti-CD3 CSANs and anti-CD133/anti-CD3 CSANs. EpCAM has been shown to be a ubiquitous marker of a number of epithelial cancers, including breast, prostate, lung, pancreatic and colon cancer, while CD133 has been shown to be a marker for a number of types of cancer stem cells. The bispecific CSANs rapidly (min) and stably (days) bind to CD3 on T-cell membranes, thus forming ***Prosthetic Antigen Receptors (PARs) T-cells***. (Upon incubation of the PAR T-cells with EpCAM+ and/or CD133+ cancer cells, T-cell activation and selective cell killing was observed. ***Furthermore, we demonstrated with murine xenograft models that the anti-EpCAM and anti-CD133 PAR T-cells are able to permanently eradicate tumors in animals.*** A unique feature of our approach is the ability to remove the CSANs from the T-cells by incubation with the FDA-approved non-toxic antibiotic trimethoprim, thus allowing us to deactivate the modified cells pharmacologically. In addition, we have shown that the CSANs have negligible immunogenicity in mice and do not induce naïve human T-cell anergy (i.e, unresponsiveness). Thus, PAR T-cells may afford and alternative and complimentary immunotherapy approach.

538021 Inhibition of Peanut Induced Mast Cell Degranulation By Designing Covalent Heterobivalent Inhibitors

Abstract:

Food allergy, Type I Hypersensitivity, is induced by abnormal immune responses from immune cells by recognizing food proteins and releasing inflammatory molecules, such as histamine and cytokines. Food allergy specific immunoglobulin E (sIgE) bound to Fc epsilon receptor I (FcÎµRI) on mast cells are crosslinked by multivalently interacting with food proteins and begin cellular degranulation which may cause detrimental immune responses such as anaphylactic shock. Several first line treatments are available for allergies such as anti-histamine, but they only prevent allergic reactions after the symptoms have occurred and induce numerous side effects, such as immunosuppression. To develop therapeutics that can cure allergic reactions, it is important to target sIgEs so that it prevents crosslinking of sIgEs by allergen protein which eventually inhibits cellular degranulation. Several efforts have been made in the past to design inhibitors regarding this aspect, but it has not been successful because the major challenges for therapies that specifically inhibit sIgEs are to identify the immunogenic epitopes presented on sIgE binding allergen proteins and offset the high sIgE-allergen affinity. Past research has identified several epitopes from allergen proteins via microarray, but these studies only evaluated the binding of epitopes to sIgE, not whether the epitopes activate sIgE-dependent immune responses. Here, we present a synthetic liposomal allergen, we call nanoallergens, that enables the identification of crucial immunogenic sIgE binding epitopes and the development of inhibitors, termed covalent heterobivalent inhibitors (cHBI). Utilizing identified immunodominant epitopes, cHBIs selectively and irreversibly bind to sIgE to prevent allergic reactions. Using sixteen peanut allergy patient sera, we evaluated the most immunodominant epitopes from major peanut proteins, Arah2 and Arah6, via nanoallergens and designed cHBI that permanently inhibits the crucial sIgEs for peanut allergies. We demonstrated a complete inhibition of degranulation responses utilizing a combination of two cHBI via *in vitro* degranulation assays and *ex vivo* basophil activation test. This study presents promising preclinical data that further development and assessment of cHBI can be potential therapeutics to prevent food allergies.

531721 Remote Control of Engineered T Cells Using Photothermal Pulses

Abstract:

Remote Control of Engineered T Cells Using Photothermal Pulses

Ian C. Miller, Marielena Gamboa Castro, Lee-Kai Sun, Jason P. Weis and Gabriel A. Kwong

Georgia Institute of Technology and Emory School of Medicine

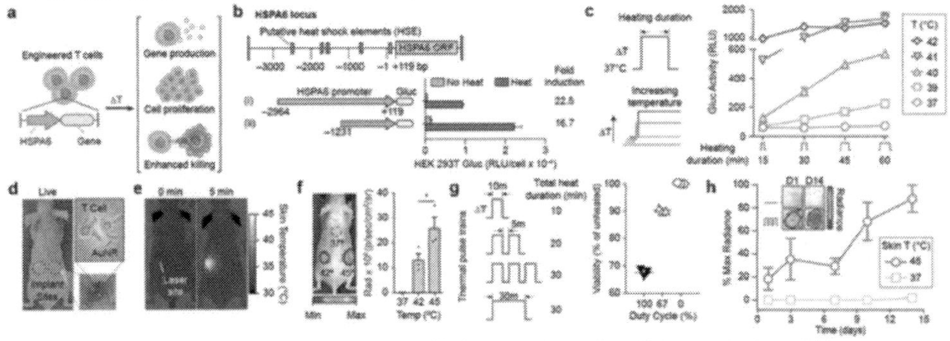

Figure 1 (a) Thermal gene switches afford control of cellular activity. (b) Two constructs cloned from the heat shock protein HSPA6 locus and used to evaluate sensitivity to thermal activation in HEK 293T cells. Fold inductions of normalized luminescence (Heat/No heat) are listed to the right of both constructs. RLU: Relative Luminescence Units, n = 3. (c) Continuous heat treatment profiles with increasing time or temperature in Jurkat T cells. Luminescent traces show that increases in both duration and temperature of heating augment switch activity, n = 3. (d) Photograph of nude mouse with subcutaneous matrigel implants (inset) containing engineered Jurkat T cells and AuNRs before heating. (e) Serial thermal images of mouse bearing AuNR-matrigel implants during laser heating. (f) Radiant image of nude mouse with Jurkat implants after heating at skin temperatures of 37, 42, and 45 °C for 20 min. Radiance calculated as the difference in value between implant site luminescence and background radiance of mouse skin, n = 3. (g) Diagram of thermal pulse trains at a 67% duty cycle (10 min on, 5 min off) and continuous heat treatments. Total heated time for last two regimens were identical (30 min). (h) Radiance trace of implant sites after pulsatile heat treatments on days 1, 3, 7, 10, and 14 after implantation, n = 4. Inset: luminescent images of representative implant sites on days 1 and 14. Error bars show SEM.

Introduction: Recent advances in synthetic biology are providing new tools to modulate cellular activity and elucidate complex signaling pathways in T cells. These include the ability to redirect T cell specificity towards cancer-associated antigens, migrate towards synthetic chemical cues, or sense combinatorial antigens using Boolean logic. Yet despite such striking progress, our ability to precisely control the activity of engineered T cells *in vivo* remains limited. This issue is a barrier to widespread application of T cell therapies in solid tumors and has raised safety concerns stemming from off-target toxicity in healthy tissues. The next generation of engineered T cells that can be remotely controlled by external cues, such as light or heat, will enhance the field's ability to dictate cellular behavior to augment anti-tumor immune activity. We leverage the precision with which heat can be spatially targeted by photothermal pulses, and engineer T cells with thermal gene switches which are constructed from the heat shock protein HSPA6 promoter (**Fig. 1a**). Using localized heating, we trigger gene expression of engineered T cells to levels 200-fold over basal activity in mice. Additionally, we show that delivery of heat as thermal pulse trains increases T cell thermal tolerance compared to continuous heating profiles with identical areas-under-the-curve (AUC). In the future, this strategy could increase the safety and potency of engineered T cell therapies for cancer.

Results and Discussion: To construct a thermal gene switch, we identified a construct of the HSPA6 promoter with low basal activity and strong heat induction, and characterized its response to mild hyperthermia (40 – 42 °C) in Jurkat T cells (**Fig. 1b,c**). We also used NIR laser light and injected gold nanorods to precisely heat targeted sites *in vivo* and increase translational activity of engineered cells greater than 200-fold (**Fig. 1d,e,f**). Heating samples using pulse trains, in contrast to continuous heating profiles, increased thermal tolerance of T cells (**Fig. 1g**) and enabled long-term remote control over cellular activity *in vivo* (**Fig. 1h**). Accordingly, this platform may be used to augment the therapeutic efficacy of engineered T cells in clinical settings.

Conclusion: Using photothermal activation of thermal bioswitches, we controlled the translational activity of engineered cells *in vivo*. Looking forward, this platform could enable precise control over cellular behavior with exquisite spatial resolution and provide an orthogonal mechanism to dictate cellular activity in addition to small-molecule or light-based methods.

536910 Serum Antibody Profiling of Nivolumab/Azacytidine-Treated Acute Myeloid Leukemia Patients Via High-Throughput Sequencing of Peptide Phage-Display Library

Abstract:

Phage panning combined with high-throughput sequencing permits profiling of humoral responses elicited by the immune system against foreign entities like pathogens, or the host's tissues in the context of autoimmune disease and cancer. Part of this response involves the production of antibodies — the detection of which forms the basis of many diagnostic tests. Conventional antibody-based diagnostic tests rely on an antigen or antigens to be known. However, for this approach to work, it implies that suitable antigenic targets are available and, more importantly, it is possible to deduce the disease or to shortlist diseases considered suspect for testing. Phage panning provides an attractive way of multiplexing a person's entire antibody repertoire by recovering the antigen-mimicking fraction of the phage display library. High-throughput sequencing allows the peptide sequence space to be deduced, establishing biomarker patterns correlated to the individual's disease state. In this study, a combinatorial library of phage-displayed linear dodecapeptides was offered to sera from an acute myeloid leukemia cohort before and after the first cycle of combination therapy (nivolumab + azacytidine). Our strategy involves incorporating two sets of samples as controls: (1) comparison of the sequence space in the naïve library before and after phage amplification to identify propagation-biased subsets and, (2) screening AML sequences against motifs derived from a flu-vaccinated cohort. Candidate peptides are matched to a custom reference database of human proteins and other human-associated microorganisms. Findings on sequence motifs and implications of custom database matches will be discussed.

537510 Yeast Surface Display Techniques Enhance Development of Chimeric Antigen Receptors for Hematologic Malignancies

Abstract:

Chimeric antigen receptor (CAR) engineered T cells have energized the field of cancer immunotherapy with their proven ability to treat $CD19^{+}$ malignancies in the clinic and emerging efficacy in treating other diseases. The synthetic CAR receptor imparts T cells with the ability to recognize antigens independent of peptide presentation by major histocompatibility complexes. The CAR antigen recognition domain often requires much engineering to generate binding interactions resulting in target-specific activation. Unfortunately, developing a functional CAR with current standard methods is empirical and inefficient, requiring individual cloning of combinations of binding domains (often scFvs) and spacer regions followed by low throughput well plate and mouse experiments to verify activity. The scFvs incorporated into CARs often derive from published antibody sequences reformatted into scFvs, which frequently results in a loss of affinity and/or stability. These antibodies are regularly of murine origin, raising the risk of immune response that may hamper treatment.

This work aims to enhance the efficiency and success of CAR development by incorporating protein engineering and combinatorial library screening using yeast surface display techniques to facilitate scFv humanization and affinity tuning, and applying the resulting CARs for treatment of hematologic malignancies.

Two computational humanization methods were applied to the antigen recognition domain of a CD19-targeted CAR previously developed in the Forman lab. Yeast surface display techniques were applied to screen expression, foldedness, and affinity of the humanized variants, adding significant efficiency to the CAR development process that generally requires lentiviral transduction of cell lines to verify these characteristics. The two humanized scFvs displayed well, but exhibited markedly diminished affinity (8 ± 1 μM and 45 ± 30 μM, respectively) compared to the murine parent (4.8 ± 0.7 nM). The humanized scFvs were formatted into CARs and expressed in human T cells. Despite the decreased affinity, the resulting CARs showed comparable activity to the equivalent murine versions in degranulation, internal cytokine staining, proliferation, and long-term killing assays against CD19+ Daudi, TM-LCL, and SUP-B15 cells. Using a systemic CD19+ NALM6 model in NSG mice, the 8 μM affinity humanized variant showed comparable therapeutic efficacy to the murine parent. Additional efforts to affinity mature these humanized variants using yeast surface display will be discussed.

In a second project, a CD123-targeted CAR developed previously in the Forman lab was humanized. The resulting clone suffered a 575-fold affinity decrease (0.26 ± 0.1 nM vs 150 ± 40 nM) relative to the murine parent. Error-prone PCR was applied to the humanized variant, and the resulting sublibrary was yeast displayed as a novel CAR-like scFv-IgG4 hinge fusion and sorted for increased or diminished binding, yielding softly mutated variants with 5.5 ± 0.2 nM, 62 ± 2 nM, and 8.6 ± 2 μM affinity. In ongoing experiments, the resulting CARs will be screened for selectivity against $CD123^{+}$ and CD123-dim cells for treatment of acute myeloid leukemia.

Overall, this work will greatly increase the efficiency of CAR development by streamlining this process using computational tools and high throughput screening. The insight that we gain from the interplay of CAR affinity, antigen density, and T cell response will be valuable in informing future CAR design.

539380 Efficient Incorporation of Matrix Protein M2 into Influenza Virus-like Particles (VLPs) for Improved Vaccine Manufacturing and Efficacy

Abstract:

Influenza virus remains a heavy economic burden and poses pandemic threats to public health worldwide, causing 250,000-500,000 deaths annually and 3-5 million cases of severe illness. Over the last 15 years in the United States, reported efficacy of flu shots has been as low as 19%. The low vaccine efficacy is primarily due to key limitations of the traditional egg-based manufacturing platform, a slow process (6-9 months) during which mutations in the circulating virus (antigenic drift) or egg-adapted vaccine virus can result in antigen mismatch. This antigen mismatch is particularly problematic for influenza strains that mutate at a faster rate, such as the H3N2 circulated in the past two flu seasons, causing higher number of hospitalizations and deaths compared to previous years. To address these issues, alternative manufacturing platforms with shorter lead times (2-3 months) such as the baculovirus-insect system have been utilized for producing recombinant protein-based vaccines (Flublok®) as well as virus-like particles (VLPs). Compared to recombinant protein-based vaccines, VLPs are more immunogenic due to greater antigen display in a three dimensional structure that resembles the influenza virus but lacks viral RNA. Furthermore, the baculovirus-insect system allows for facile engineering of the antigen composition of the VLPs, which is critical for developing broadly protective flu vaccines and ultimately eliminating the current practice of updating flu shots annually necessitated by antigen mismatch.

Towards the goal of rapid manufacturing of broadly protective influenza vaccines, VLPs presenting antigens that are conserved among multiple influenza strains such as the matrix protein M2 have been attempted. However, the expression of full-length M2 showed cytopathic effects on insect cells hindering their ability to produce M2-containing VLPs. In this work, we quantified the cytopathic effects of expressing M2 in insect cells using various metrics and revealed that M2 significantly impedes the ability of insect cells to replicate baculovirus but does not directly cause cell death. Because baculovirus is the gene carrier for expressing influenza viral proteins including M2, HA and M1 in insect cells, the reduced baculovirus replication leads to reduced expression of all three influenza proteins, which in turn reduces the quantity and quality (i.e., antigen density) of the resulting VLPs. Based on the function of M2, we hypothesized that blocking its ion channel activity could address this issue. Indeed, the use of an M2 inhibitor restored the baculovirus replication when expressing M2 in insect cells, resulting in > 5-fold improvement in the expression of all three influenza viral proteins M2, HA and M1. More importantly, it further improved both the quantity and the quality of the VLPs produced, as evidenced by the increased number and antigen (HA) density of the VLPs. For future work, we will evaluate the vaccination efficacy of the M2-incorporated VLPs in mice. Given the importance of HA density and the ability of M2 ectodomain in eliciting effective, broadly protective and long lasting immunity against influenza, this work provides a novel approach for manufacturing and engineering effective and broadly protective influenza vaccines.

540761 Cytokine-like Regulation of T-Cell Fate Mediated By Tryptophan-Derived Microbiota Metabolites

Abstract:

Investigating the functional role of the microbiota in human health and its regulation of the immune system is one of the most exciting and impactful areas of biomedical research. However, despite the undeniable link between the microbiota and human health, the mechanisms used by the microbiota to mediate beneficial and deleterious effects on host immunity are poorly understood. An emerging mechanistic paradigm is that communication of the microbiota with host immunity is mediated via distinct non-host derived metabolites, such as short chain fatty acids. Prior work from our labs demonstrated that tryptophan- (Trp)-derived metabolites are an abundant and novel class of immunomodulatory microbiota signals that regulate inflammation as well as are novel endogenous ligands of the aryl hydrocarbon receptor (AhR) (Sridharan et al., 2014; Jin et al., 2014). With the growing number of pivotal roles that the AhR has for host immunity, particularly in the gastrointestinal (GI) tract, we hypothesized that microbiota-derived Trp-metabolites are important mediators of GI tract mucosal immunity, in particular, CD4+ and CD8+ T-cell function. We screened a panel of six Trp metabolites for their ability to regulate T-cell differentiation. Of these, two metabolites - indole and 5-hydroxyindole (5-HI) - exhibited the strongest and most interesting properties. We determined that indole promoted the differentiation of naïve T-cells to Tregs, both in the absence and presence of Treg skew cytokines. At the same time, indole also attenuated the differentiation of naïve T-cells to Th17 cells under Th17 skew conditions. In contrast, 5-HI inhibited Tregs while promoting Th17 and Th1 development and function. In co-conditioning experiments, the T-cell instructing activities of indole were dominant over 5-HI. Since both indole and 5-HI are bona fide ligands for the AhR, we used $AhR^{-/-}$ mice to determine that indole's effects on T-cell differentiation are AhR-dependent, while 5-HI's effects are largely AhR-independent. A similar reciprocal regulation by indole and 5-HI was observed with human T-cells. Our work shows that two microbiota-produced small molecules, indole and 5-hydroxyindole, have reciprocal and contrasting roles for regulating T-lymphocyte fate and function and indicate novel "cytokine-like properties" for these metabolites.

Funding: This is supported by the NIH 1R01 AI110642-01

532354 Macrophage Checkpoint Blockade in a Cell-Based Immunotherapy Can Generate Durable and Safe Cures in a Poorly Immunogenic, Syngeneic Mouse Tumor Model

Abstract:

Bone marrow-derived monocytes and macrophages can be engineered to phagocytose opsonized cancer cells by blockade of the inhibitory surface receptor SIRPα which recognizes the "self-marker" CD47 expressed on all cells. Here, we explore this strategy in a cellular immunotherapy targeting syngeneic, orthotopic B16 melanoma in immunocompetent mice. These tumors double in size every two to three days, which is ~100-fold faster than human tumors, but are difficult to treat and regarded as poorly immunogenic, making them refractory to many current immunotherapies including blockade of CD47 with injected antibodies. We harvested marrow cells from donor mice and incubated with (1) anti-SIRPα to block the anti-phagocytic, "don't eat me" signal, and (2) anti-Tyrp1 against tyrosinase-related-protein-1 on B16 cells. The latter antibody binds Fc-receptors on the marrow monocytes, targeting them to tumors and providing a pro-phagocytic, "eat me" signal. Tail-vein injection of the engineered cells into mice inoculated with B16 cells can prevent tumor growth and result in durable cures, with some cured mice developing long-term adaptive immunity to protect against re-challenge. The approach also avoids safety concerns including anemia that are already evident in the clinic with intravenous anti-CD47 blockade.

517859 Development of a Kynureninase Clinical Candidate, a First-in-Class Enzymatic Checkpoint Inhibitor

Abstract:

Tumors inhibit immune responses largely through two mechanisms: (1) Through protein-protein interactions (e.g. PD1:PD-L1) and (2) By cancer-induced accumulation of immunosuppressive metabolites in the tumor microenvironment. The most potent such metabolite is the tryptophan-oxidation product kynurenine (Kyn) that induces tolerance through many mechanisms (e.g. promotes Treg differentiation and inhibits CD8+ T cell function) and promotes tumor survival and motility. Kyn is synthesized *in vivo* by three different enzymes: IDO1, IDO2, and TDO. Pharmacological blockade of the Kyn pathway is highly sought after, and three small molecule inhibitors of IDO1 are in clinical trials with TDO inhibitors expected to enter the clinic soon. However, because of the redundancy of the Kyn synthesis pathway, inhibition of IDO1 or TDO alone cannot eliminate Kyn accumulation, a fact underscored by the modest effects reported for inhibitors to date. Additionally, complete blockade of Kyn synthesis would disrupt intracellular Kyn homeostasis leading to toxicities (through loss of NAD synthesis etc.). Fortuitously, excess Kyn must be physically exported out of tumor cells and into immune cells in order to have an immunosuppressive effect, giving the opportunity to prevent Kyn immune suppression by eliminating it from the tumor microenvironment and serum. Therefore, we resolved to eliminate the extracellular Kyn pool though a completely different mechanism - the systemic administration of a Kyn-degrading enzyme, kynureninase.

As a proof of concept, we utilized a highly active bacterial kynureninase to demonstrate that enzymatic elimination of tumor-produced Kyn is an effective anti-cancer therapy. Administration of the bacterial kynureninase to B16-OVA melanoma syngrafts in C57BL/6J mice reduced serum Kyn level, resulted in significant tumor growth retardation, and extended survival in a manner indistinguishable from that observed with the immune checkpoint inhibitor antibodies, Î±-PD-1 or Î±-CTLA-4. Despite its effectiveness, the bacterial kynureninase is not a realistic therapy for clinical use because it is highly immunogenic and has low serum stability ($T_{1/2}$ = ~1 hour). There exists a human kynureninase enzyme, but it is ~700X weaker than the bacterial enzyme so it elicits no *in vivo* anti-tumor effect. Therefore, we have undertaken a large-scale protein engineering effort to increase the catalytic active and stability of the non-immunogenic, human kynureninase enzyme to develop a clinical candidate. We first developed parallel genetic selection systems in *E. coli* and *S. cerevisiae* that enable high throughput screening of >10^8 variants per library. We then performed numerous rounds of directed evolution, employing error-prone library construction, comprehensive codon mutagenesis, targeted mutagenesis of disparate residues, phylogenetic guided mutations, and DNA shuffling. Through a step-wise process of improving enzyme activity, introducing key substrate docking residues important for substrate binding, and then further directed evolution, we successfully engineered human variants having >600X improved activity over wildtype that show *in vivo* efficacy. Further efforts to reduce mutational burden, improve stability, and optimize expression have enabled development of kynureninase variants with clinical potential.

535876 Immunotherapeutic Targeting of Cancer Cell-Associated Polysialic Acid

Abstract:

The glycocalyx consists of a dense meshwork of sugar polymers called glycans that are interspersed with transmembrane proteins and other cell surface machinery. Glycans are believed to encode the structural organization of the glycocalyx, and in doing so, tune how these systems perceive and process critical information from the microenvironment. Moreover, the glycocalyx undergoes changes with the onset of cancer and the altered glycans play functional roles in tumor growth and metastasis. One example is polysialic acid (polySia), a polymer of Î±2, 8-linked sialic acid residues that accumulates on the surface of different cancers including breast, lung, neuroblastoma, and pancreatic. In addition to functioning as a diagnostic indicator, polySia also impacts the biology of the affected cells by inhibiting adhesion and promoting metastasis. To better understand the important roles played by polySia in cancer, we are developing highly specific glycan-binding reagents for imaging studies as well as for potential therapeutic intervention. This presentation will discuss our efforts to develop and characterize polySia-specific monoclonal, Fab and scFv antibodies, including glycan microarray profiling of binding specificity, Biacore analysis of binding affinity, internalization assays to measure antibody-mediated receptor endocytosis, and evaluation of antibody-mediated tumor cell killing capabilities. Collectively, our results are helping to delineate the connection between aberrant polySia glycosylation and the invasive and metastatic potential of different cancers, as well as revealing the potential of high-affinity polySia antibodies for therapeutic use. This work contributes to a growing body of evidence that demonstrates the importance of investigating carbohydrate epitopes as targets for immunotherapy.

538162 Novel Immunomodulatory Peptide Polymers for Vascularized Composite Allograft Rejection Prevention

Abstract:

Novel Immunomodulatory Peptide Polymers for Vascularized Composite Allograft Rejection Prevention

Since the first hand transplant was performed nearly two decades ago, the use of vascularized composite allografts (VCA) provides patients who are suffering from devastating tissue loss such as limb amputation or significant facial disfigurement with full functional equivalents greatly improving their quality of life. To prevent VCA rejection, aggressive immunosuppressive strategies are employed for life which limits the patientâ€™s capacity to fight infection and undergo would healing as well as increasing their cancer risk. Unfortunately, even with this approach, approximately 55% of face transplant recipients and as many as 85% of hand transplant recipients experienced acute rejection episodes in the first year post-transplantation.

The goal of our research is to create a novel system capable of facilitating localized immunosuppression through the recruitment and polarization of a variety of immune cells (*i.e.* macrophages, dendritic cells, and T cells) towards anti-inflammatory and/or tolerogenic phenotypes. Previous research has determined that the hormone-related vasoactive intestinal peptide (VIP) possesses potent pleotropic effects including strong immunomodulatory capacity. Unfortunately, VIP has a half-life of only a few minutes in the body. To enhance the prolonged regional supply of VIP, we have created a novel degradable polyester for which VIP can be readily incorporated into the polymer backbone. Our data demonstrate that directed delivery of VIP have an immunosuppressive effect on multiple types of antigen presenting cells as well as enhance CCL22 production from dendritic cells. As CCL22 is a chemokine with strong regulatory T cell recruitment capacity, this can complement the localized anti-inflammatory effects to further prevent tissue rejection. Taken together, these results provide considerable evidence that VIP-containing polymers hold tremendous potential for limiting transplant rejection.

538557 CO-Delivery of Paclitaxel and Imatinib By PEG Derivatized NLG Carrier As Enhanced Immunochemotherapy

Abstract:

Indoleamine 2,3-dioxygenase (IDO), which catalyzes the initial and rate-limiting step in the tryptophan catabolism along the kynurenine pathway, is an immunosuppressive enzyme that can lead to T cell anergy, apoptosis and potential induction of regulatory T cells (T_{reg} cells). The consideration of IDO has driven the development of several IDO inhibitors including NLG919, an immunotherapy drug that targets IDO with high efficacy, into the clinical trial for treating solid tumor. Immunochemotherapy that combines the chemotherapeutic agents with the immune checkpoint inhibitors represents one of attractive strategies for cancer treatment. Previously, our group has developed a PEG_{2K}-Fmoc-NLG micellar system that can effectively deliver anticancer drug Paclitaxel (PTX) and showed enhanced effect with chemotherapy to inhibit tumor growth. NLG is known as an enzymatic inhibitor of IDO, but due to the feedback mechanism, the decrease of enzymatic metabolites would lead to the up-regulation of IDO expression. For long-term treatment, increased dose of this formulation would be necessary to maintain the same effect for treating cancer. Meanwhile, we also found PTX itself could up-regulate IDO mRNA expression. Hence we screened out another anti-cancer drug Imatinib (Gleevec®) which shows a down-regulation effect on IDO expression and developed a PEG_{5K}-$(Fmoc\text{-}NLG)_2$ prodrug carrier for effective co-delivery of Paclitaxel and Imatinib.

In this study, we have developed a simple and dual-functional nanomicellar system that consists of NLG-based hydrophobic domain, a PEG hydrophilic segment and a drug-interactive Fmoc motif. In addition to significantly enhancement of drug loading capacity and formulation stability, our improved PEG_{5K}-$(Fmoc\text{-}NLG)_2$ system also ensured the effective co-delivery of the prodrug PEG-NLG and two anti-cancer agents PTX and IMA to the targeted tumor site. The rapid release of PTX could mediate the killing of tumor cells and/or cause direct effect on immune cells, which resulted in enhanced antigen presentation that lead to immune response. Meanwhile, the slow release of NLG919 from the prodrug facilitated the immune response by preventing or slowing down the development of IDO-induced immune escape. More importantly, the release of IMA blocked IDO at the level of gene expression, which further sustained or enhanced the magnitude of immune response. No synergism has been found in the combination of PTX and IMA in inhibiting tumor cell proliferation *in vitro*. However, we observed an enhanced anti-tumor activity of PTX+IMA loaded PEG_{5k}-$(Fmoc\text{-}NLG)_2$ micellar system in an immunocompetent mouse model, which suggests that the efficacy of IMA might come from its immune-stimulatory effect. Indeed, it seems to be bright for combining IDO inhibitors and targeted drugs, which can interfere with immune escape pathways as enhanced immunochemotherapy.

541021 Challenges in Continuous Manufacturing of Biologics: Summary of NASEM Workshop

Abstract:

The National Academies of Sciences, Engineering and Medicineâ€™s Board of Chemical Sciences and Technology hosted a workshop on the above theme on July 30-31, 2018 under the sponsorship of the USDA FDA and BARDA. This presentation will present a summary of the factors and issues underlying the business case for going continuous as well as specific technical challenges identified in upstream processing, downstream processing, product manufacture, integration, and validation. The content of this presentation will draw from the report of the workshop that will be issued late in 2018.

541014 Leveraging Biomaterial Stability and Retention in Lymph Nodes to Control Immunity

Abstract:

Lymph nodes are key tissues that coordinate immune function. In this talk I will present some of the new ways we are using engineered materials to study the local microenvironment of lymph nodes to inform design of vaccines and immunotherapies. Many common biomaterials exhibit intrinsic physiochemical features that activate, suppress, or modulate immune pathways, even in the absence of other immune signals. In this area, I will discuss our work to characterize how these immunological interactions change as common polymeric vaccine carriers degrade in immune cells and in lymph nodes. Understanding how the immunological profile of biomaterials evolves during delivery could enable more rational design of new vaccines and immunotherapies. In a second area, I will discuss our work to understand how the local concentrations of drugs in lymph nodes impacts the development of immune cells expanding and differentiation in these sites. In particular, our data reveal that different types of T cell responses â€“ effector, memory, regulatory â€“ can be produced against the same antigen by locally altering metabolism in lymph nodes during T cell expansion. As translational examples, I will discuss our work to exploit this idea using immune-modulatory small molecules to enhance responses for cancer vaccination, or conversely, to combat autoimmunity by regulating T cell function.

541013 Improving Homogeneous Differentiation of Stem Cells By CRISPR-Mediated Knockout

Abstract:

Induced differentiation of stem cells in culture can result in heterogeneous mixtures of cells contaminated with off-target cell types, a reflection of intrinsic heterogeneity within the stem cell population or stochastic events. This hinders efficiency of stem cell differentiation *in vitro* by requiring extra selection and sorting protocols. Control of cell fate during differentiation is strongly influenced by the activity of transcription factors. We hypothesized that we could eliminate generation of off-target cell types by controlling the availability of transcription factors.

Previously, using forskolin and IBMX to induce dopaminergic differentiation of mesenchymal stem cells, we observed a small population of glutamate sensitive cells indicating heterogeneous stem cell differentiation. We hypothesized that by creating mesenchymal stem cell lines null for transcription factors specifying the glutamatergic phenotype we could produce a more homogeneous mixture of dopaminergic-like cells. We show that MSCs exhibit heterogeneous differentiation during dopaminergic differentiation and aim to create Tbr1$^{â€“/â€“}$ MSCs that could potentially yield a more homogeneous dopaminergic population after induction with forskolin and IBMX.

541007 Liposome Targeted Proliferation of Cytotoxic T Cells

Abstract:

Adoptive cell therapy (ACT) is an evolving personalized cancer immunotherapy. The approach has shown promising results in treating metastatic melanomas but has limited success in other malignancies due to: 1) lack of cytotoxic T cells found within tumor mileau, 2) the difficulty and cost associated with genetically modifying and expanding small populations of T cells ex vivo to therapeutic concentrations and 3) short term persistence of the inoculated T cells. To address these limitations, we hypothesized that liposomes can be used to deliver the T cell growth factor 'IL-2' directly to the cytotoxic T cells within the body, thereby circumventing the need for ex vivo expansion and significantly reducing the associated costs and complexities. Towards this goal, we have developed and characterized a novel approach to synthesize small unilamellar liposomes encapsulated with the desired molecules at high concentrations. We have also developed a unique thermal equilibration technique to co-load both hydrophobic and hydrophilic compounds within these liposomes at high concentrations. Utilizing this technique, liposomes containing IL-2 were created and functionalized with anti-CD8, and delivered to Tall-104 model cell line in vitro to demonstrate the efficacy of liposomal IL-2 in their proliferation and cytotoxicity against breast cancer cells.

538217 Immunomodulatory Biomaterials: The Quest for Fundamental Design Rules

Abstract:

Research over the past few decades has revealed the incredible potential biomaterials possess to be utilized for immunological applications. Whether they function as traditional vaccines, cancer immunotherapeutics, autoimmune disease treatments, or transplant rejection prevention medications, immunomodulatory biomaterials have made tremendous impacts in both the laboratory and the clinic. While exciting advancements have been made, significant effort has focused on parsing out minute differences among materials platforms instead of concentrating on the underlying phenomena that govern materials immunology. Through discussing some recent efforts in this area, I hope to highlight the potential we have as a community of immunomodulatory biomaterials researchers to come together to uncover fundamental design rules that will dramatically enhance our scientific and clinical impact.

530275 Does Co-Encapsulation Matter?: Probing the Biophysical and Functional Impacts of Nanoparticle Combinatorial Delivery

Abstract:

One clear benefit of using nanoparticles (NPs) for drug delivery and immunotherapy for cancer and autoimmune disorders is the designable, predictable manner of achieving combination delivery. However, an outstanding question is whether multi-payload combination therapy using NPs is best achieved by either a) Separately-encapsulated, or b) Co-encapsulated NPs. We address this question of encapsulation strategies by probing the biophysical implications of delivering differentially-encapsulated NPs, utilizing dye-loaded NPs and *in silico* modeling of NP uptake. Further, functional and therapeutic impacts were investigated focusing on models of antigen uptake and presentation, employing Separately- or Co- NP systems of antigen delivered with pro- or anti-inflammatory (tolerogenic) adjuvants. In the process of this investigation, we discovered that encapsulation strategies in NPs play a critical role in the spatio-temporal biodistribution kinetics of NPs in target cells and, thus, profoundly influence therapeutic outcomes. Our results provide biophysical and therapeutic insights into the significance of encapsulation strategies that can guide the designs of future NP-mediated combination therapies.

536131 Pollen Grains - a Novel Biomaterial for Oral Vaccination

Abstract:

BACKGROUND: Oral vaccination is one of the â€˜holy grailsâ€™ in the field of vaccine delivery. We have recently shown that pollen grains, the bane of allergies for some, can be repurposed and engineered to function as microcapsules and microcontainers for transport of vaccines across the harsh environment of the stomach. The unique feature of pollens is that their shell is made up of a resistant material that can withstand the acidic conditions of the stomach without degradation. Here we studied different aspects of pollens as a novel biomaterial for oral vaccination.

METHODS: Ragweed pollen were chemically treated to strip them off their cellular material and biomolecules. Pollens were then characterized using elemental analysis, SDS-PAGE, and scanning electron microscopy (SEM). Biocompatibility and safety was evaluated on Caco-2 cell monolayers, which served as a model intestinal epithelial cell line. Immunomodulatory properties of ragweed pollens were evaluated by culturing bone-marrow derived macrophage and dendritic cells in the presence of ragweed pollens and quantifying secreted pro-inflammatory cytokines and cell surface expression markers. The ability of ragweed pollens to generate systemic and mucosal antibody responses was evaluated in mice by orally vaccinating Balb/c mice with ovalbumin as a model antigen formulated with ragweed pollens.

RESULTS: SEM micrographs showed that chemical treatment of pollens helped to produce pollens that were intact and clean. Nitrogen analysis coupled with SDS-PAGE further confirmed that no protein was detectable in the pollens after treatment. Biocompatibility and safety studies on Caco-2 cells showed that the electrical resistance across the epithelial cells did not reduce upon addition of different concentrations of pollens. Furthermore, as compared to poly-lactic-co-glycolic acid particles, no significant increase in cytotoxicity was observed. Macrophages and dendritic cells released pro-inflammatory cytokines and also upregulated surface markers of activation. This shows that pollen grains have an immunomodulatory property. Indeed, when pollens were used in vivo for oral vaccination, a significant increase in systemic IgG and mucosal IgA antibodies against ovalbumin (model antigen) were seen as compared to ovalbumin alone cohorts. Examination of IgE antibody in the mouse serum against ragweed extract did not show any increase indicating that ragweed pollens, which were stripped and cleaned off their natural ragweed allergens, do not induce an allergic response. Lastly, ex vivo examination of mouse intestinal tissue via confocal microscopy showed that ragweed pollens can penetrate the intestinal epithelial barrier and reach the sub-mucosal space.

CONCLUSIONS: The data of this study provides proof-of-concept that pollen grains can be repurposed for use as a natural biomaterial for oral vaccination. This data also sheds light on the immunomodulatory property of pollens and offers insight into the potential mechanism of how pollen grains help to generate an immune response. The data thus supports future development of pollens for oral vaccine delivery.

534081 Mucosal Polyanhydride Nanovaccine Against Respiratory Syncytial Virus Infection in the Neonatal Calf

Abstract:

Human respiratory syncytial virus is a leading cause of lower respiratory tract disease in infants and young children worldwide, with an estimated 33 million new cases annually in children under five years of age and more than 100,000 resultant deaths. Bovine RSV (BRSV) is closely related to HRSV and a significant cause of morbidity in young cattle. BRSV infection in calves displays many similarities to RSV infection in humans, including similar age dependency and disease pathogenesis, making it an attractive animal model for vaccine studies with HRSV. BRSV is an enveloped, negative-stranded RNA virus that whose viral envelope is lipid derived from the host plasma membrane, but also contains three virally encoded transmembrane surface glycoproteins, namely the large major attachment glycoprotein G, the type I viral fusion protein F, and small hydrophobic protein SH, the former two playing a critical role in viral replication. Current vaccination strategies incorporate the F and G proteins, as these antigens contain dominant epitopes that can induce neutralizing antibodies, thus inhibiting viral attachment.

Polyanhydride nanoparticle-based vaccines (i.e., nanovaccines) represent a safe and efficacious platform technology with the capability to enhance immune responses against pathogens such as BRSV. These materials are biocompatible, providing mild inflammation at the injection site. Due to their hydrophobic properties, they provide sustained release of encapsulated proteins, which allows for dose sparing, and display chemistry-dependent internalization and activation by antigen presenting cells, which allows for the ability to modulate immune responses. In the current study, the efficacy of a polyanhydride nanovaccine encapsulating the BRSV F and G glycoproteins was evaluated in the neonatal calf. The nanovaccine released immunologically active antigen, evidenced by the ability of BRSV immune serum to bind the released proteins, as well as stimulate CD4$^+$ T cells from BRSV-immune cows. The nanovaccine demonstrated the ability to activate bovine monocyte-derived dendritic cells and alveolar macrophages through the secretion of IFN-Æ”, TNF-Î±, IL-6, and IL-1Î². Following intranasal administration to calves, the nanovaccine allowed for reduced gross pathology and histopathology in the lungs as well as viral burden in the lungs. Nasal fluid and bronchoalveolar lavage fluid showed statistically significantly higher levels of IgA compared to unvaccinated controls, as well as higher viral neutralization titer six days post-challenge. Peripheral blood mononuclear cells from the nanovaccine-immunized calves stimulated with the F and G proteins demonstrated enhanced T cell responses compared to unvaccinated calves, with a significantly higher degree of dividing T cells, as well as secreted IFN-Æ” and IL-17. These results demonstrate that a single intranasal dose of a polyanhydride nanovaccine encapsulating the BRSV F and G glycoproteins was an effective vaccine platform that allowed for robust adaptive immune responses against BRSV.

534987 Local Induction of Endogenous Regulatory T Cells for the Treatment of Periodontal Disease

Abstract:

Periodontitis (periodontal disease) is one of the most prevalent diseases today, affecting over 65 million adults in the United States alone. The disease is characterized by an overburden of invasive bacteria, gum inflammation and plaque buildup. Over time, these symptoms result in severe loss of gingival tissue attachment, tooth loss and bone resorption. Periodontitis is also correlated to higher incidences of respiratory disease, cardiovascular disease and even diabetes. Although current treatment targets the oral bacterial load through the administration of local antibiotics and procedures such as scaling and root planing, they do not address the root cause of the disease. Specifically, more recent studies suggest that although overburden of invasive bacteria is what initiates the disease, it is an unbalanced inflammatory response that propagates disease symptoms. Interestingly, in the healthy steady state, the body naturally combats destructive inflammation and imbalanced inflammatory responses through regulatory pathways mediated by cells such as regulatory T-cells (Tregs). Consequently, we have sought to target this underlying immune imbalance through increasing the local presence of regulatory lymphocytes (Tregs) in order to restore local, immunological homeostasis. Previously, we demonstrated that the presence of 3 factors ; IL-2 (cytokine- promotes the differentiation of immature T cells into regulatory T cells), Rapamycin (small molecule- mTorr inhibitor to prevent differentiation into to Tcell types) and TGF-Î² (cytokine- growth factor essential for Treg differentiation) robustly induces the transformation of naÃ¯ve CD4+ lymphocytes into FoxP3+, suppressive Tregs. Accordingly, we hypothesized that utilizing a local delivery of Treg-inducing factors could effectively treat the immune-mediated destruction observed in periodontal disease by increasing the local number of Tregs. Accordingly, we formulated 3 separate controlled release systems that can be locally placed in the periodontal space as a depot (degradable microspheres composed of polymers that have an excellent track record of FDA approval). Using the periodontopathogen *Porphyromonas gingivalis* (*Pg*) to induce periodontal disease in a Balb/c mouse model, we have demonstrated that this combination of factors significantly decreases alveolar bone loss in mice. A shift in microenvironment has also been demonstrated with an analysis of pro-inflammatory and anti-inflammatory markers. Ultimately, the use of these Treg-inducing microparticles significantly improves periodontitis outcomes and may be able to serve as a platform delivery system to treat other inflammatory diseases.

534294 Hydrogel-Based Cell Culture System for Scalable Expansion of Human Primary T Cells

Abstract:

Introduction: Adoptive immunotherapy refers to the transfer of immune cells (e.g. T lymphocytes) with antitumor activity into a patient to mediate tumor regression. Basic, translational and clinical studies have shown adoptive immunotherapy is highly effective for treating many cancers, such as melanoma, cervical cancer, lymphoma, leukemia. However, the cost for manufacturing T cells with current cell culturing technologies is extremely high. For instance, one dose of a recently approved engineered T cells for treating children and young adults with B-cell acute lymphoblastic leukemia costs $475,000. Thus novel cell culturing technologies that can significantly reduce the manufacturing cost and increase the manufacturing capacity is highly wanted for the widespread application of adoptive immunotherapy.

Materials and Methods: A micro-extruder was designed and made for processing alginate hydrogel tubes. To process the alginate hydrogel tubes, a solution containing single T cells, T cell activators (e.g. anti-CD3/CD28/CD2 antibodies) and 2% hyaluronic acid polymer, and a solution containing 1.5% alginate polymer was pumped into the central channel and side channel of the micro-extruder, respectively, to form a coaxial core-shell flow that was extruded into a $CaCl_2$ buffer. The shell alginate flow was instantly crosslinked by Ca^{2+} ions to form an alginate hydrogel tube. Subsequently, the $CaCl_2$ buffer was replaced by the T cell culture medium and cells were grown in the tubes. To collect or passage cells, the alginate hydrogel tubes could be dissolved with the cell-compatible ethylenediaminetetraacetic acid (EDTA) solution (0.5 mM, 5 minutes at room temperature) to release the micro cell masses that could be dissociated into single cells by gently mechanical pipetting for the following analysis or passage.

Results: Here, we report a novel cell culture technology for expanding primary human T cells for adoptive immunotherapy. T cells are suspended and cultured in microscale alginate hydrogel tubes (AlgTubes) that are suspended in the cell culture medium in a culture vessel. The hydrogel tubes protect cells from hydrodynamic stresses and confine the cell mass less than 400 Î¼m (in radial diameter) to ensure efficient mass transport, creating a cell-friendly microenvironment for growing T cells. This system is simple, scalable, highly efficient, defined, cost-effective and compatible with the current Good Manufacturing Practices. Under optimized culture conditions, the AlgTubes enable culturing T cells with high cell viability, low DNA damage, high growth rate (~320-fold expansion over 14 days), high purity (~98% CD3+) and high yield (~$3.2x10^8$ cells/mL hydrogel), all offer considerable advantages compared to current T cell culturing approaches. This new culture technology can significantly reduce the culture volume, time and cost, while increase the production capacity to make the adoptive immunotherapy broadly available and affordable.

Conclusions: In summary, we believe the alginate hydrogel tubes will significantly advance the adoptive immunotherapy and will be of broad interest to individual laboratories, institutions, and biotechnology companies working on adoptive immunotherapy.

530220 Transforming Immunotherapy with Nature-Inspired Engineering

Abstract:

In adoptive T cell therapy, T cells are extracted from cancer patients, processed *ex vivo* and then reinfused. A critical step in the processing, T cell activation, is required to promote cell expansion, differentiation and secretion of cytokines, which all contribute to the efficacy of the therapy. The current gold standard approach for *ex vivo* T cell activation involves stimulating the cells with antibody-coated microbeads. However, this approach is laborious as the beads are non-biodegradable and need to be separated before cell infusion. Furthermore, the design of these materials has overlooked the importance of biophysical parameters which could affect expansion rates and the functionality of cell products. In particular, the T cell receptor (TCR), known for receiving activation signals, is increasingly recognised as a mechanosensor that converts mechanical cues into intracellular biochemical signals. Gaining the ability to harness T cell mechanobiology may therefore provide new ways of fine-tuning the cells for therapy. Taking inspiration from cellular mechano-sensing, we investigated the regulation of T cell activation by substrate stiffness, which has been shown to influence many processes (e.g. differentiation and proliferation) in other cell types. As a proof-of-concept experiment, antibody-coated hydrogels were employed as a stiffness-tunable culture platform to stimulate Jurkat T cells. The stiffness of the hydrogels was varied by controlling the monomer-to-crosslinker ratio. Results showed differential secretion of Interleukin-2 (IL-2) from T cells stimulated on substrates of varying stiffness, suggesting that activation can be mechanically modulated. These findings further our pursuit of designing a material that is both biophysically and biochemically optimised for efficient T cell activation. Moreover, in order to introduce automation and the concept of device modularity to *ex vivo* cell processing, the hydrogel was integrated with a microfluidic platform. We anticipate that this approach can potentially set the foundations for a new class of T cell processing technology, whereby cell products can be optimised in well-controlled microenvironments for cancer immunotherapy.

AUTHOR INDEX